DATE DUE

MEASURE IT

Patty Whitehouse

Rourke
Publishing LLC
Vero Beach, Florida 32964

www.rourkepublishing.com

PHOTO CREDITS: P4, 5, 12, 13, 14, 15, 16 © Armentrout; P6, 9 © PIR; P10, 11, 20 © Craig lopetz; P17 © Lynn Stone; P7 © Anna Abejon; P8 © Jeffery Smith; P18 © Paul Cowen; P19 © Tomasz Resiak; P22 © Piotr Przeszlo

Editor: Robert Stengard-Olliges

Cover design by Nicola Stratford

Library of Congress Cataloging-in-Publication Data

Whitehouse, Patricia, 1958-
Measure it / Patty Whitehouse.
p. cm. -- (Tool kit)
Includes index.
ISBN 1-60044-209-9 (hardcover)
ISBN 1-59515-563-5 (softcover)
1. Measuring instruments--Juvenile literature. I. Title. II. Series:
Whitehouse, Patricia, 1958- Tool kit.

TJ1313.W56 2007
681'.2--dc22

2006010734

Printed in the USA

CG/CG

www.rourkepublishing.com – sales@rourkepublishing.com
Post Office Box 3328, Vero Beach, FL 32964

Table of Contents

Measuring Things

Workers are building a house. How much wood will they need? Will the windows fit?

The workers need to know the size of things. They need to measure them with **tools**.

Work and Tools

Tools help with many kinds of work. Some tools are for building. Some are for fixing things.

The tools in this book help workers measure things.

How Long?
Rulers and Tape Measures

A ruler helps you measure **length**. It is good for measuring things that are straight.

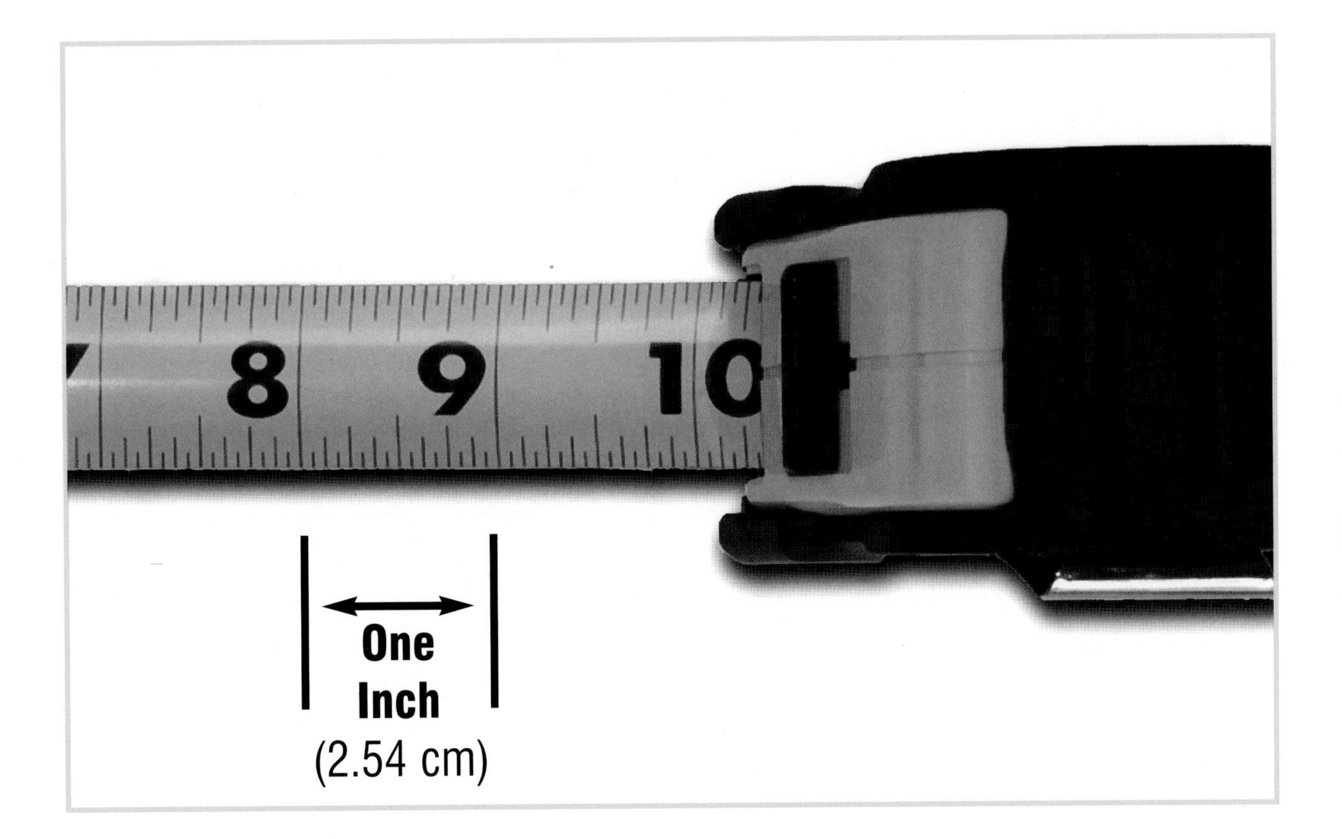

A tape measure is a ruler that rolls up. It can measure things that are longer than a ruler.

How Far?
Measuring Wheels

This is a measuring wheel. It measures **distance**. It rolls on the ground.

A measuring wheel rolls on grass, sand, or dirt. It measures straight lines. It measures curvy lines, too.

Are The Corners Right? Speed Squares

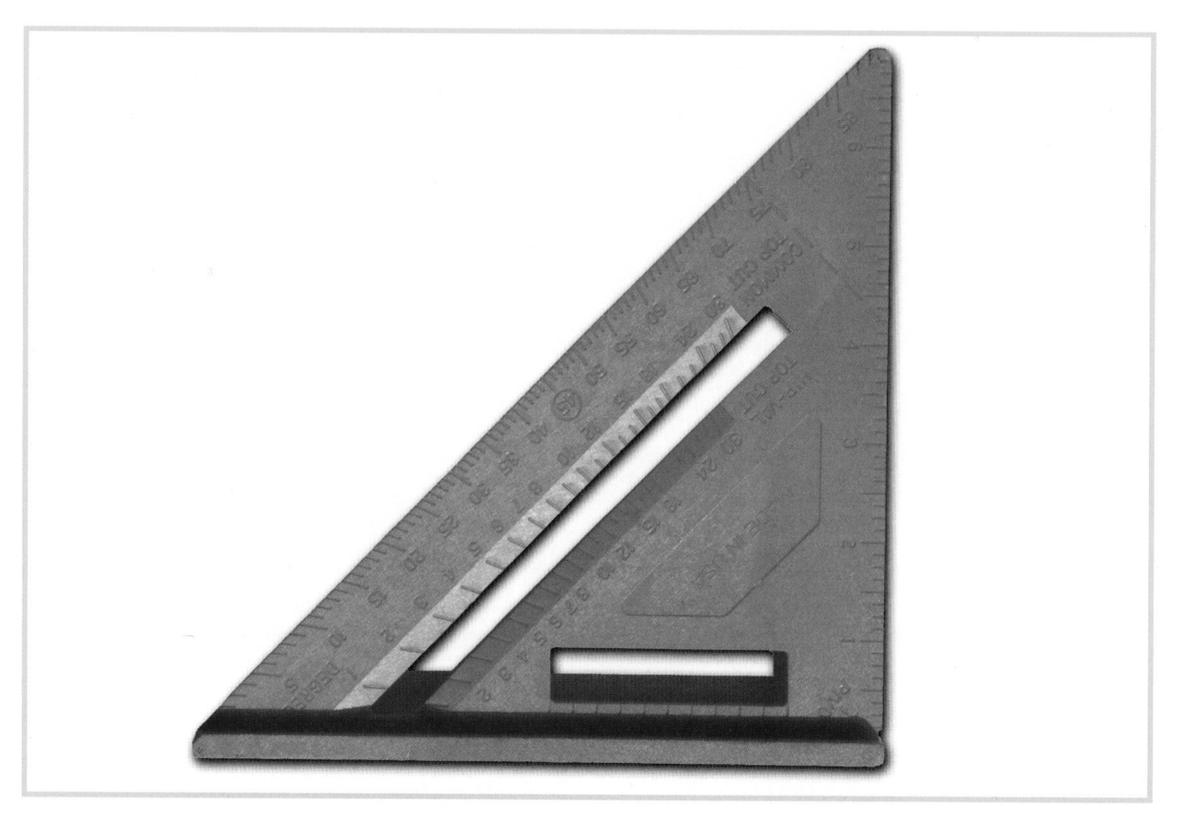

This is a speed square. **Carpenters** use it to make sure corners have an L-shape.

Windows and doors have corners. Each corner must make an L-shape. Then windows and doors will close tightly.

Is It Flat? Spirit Levels

Carpenters use a spirit **level**. Levels have three tubes with a bubble in each one.

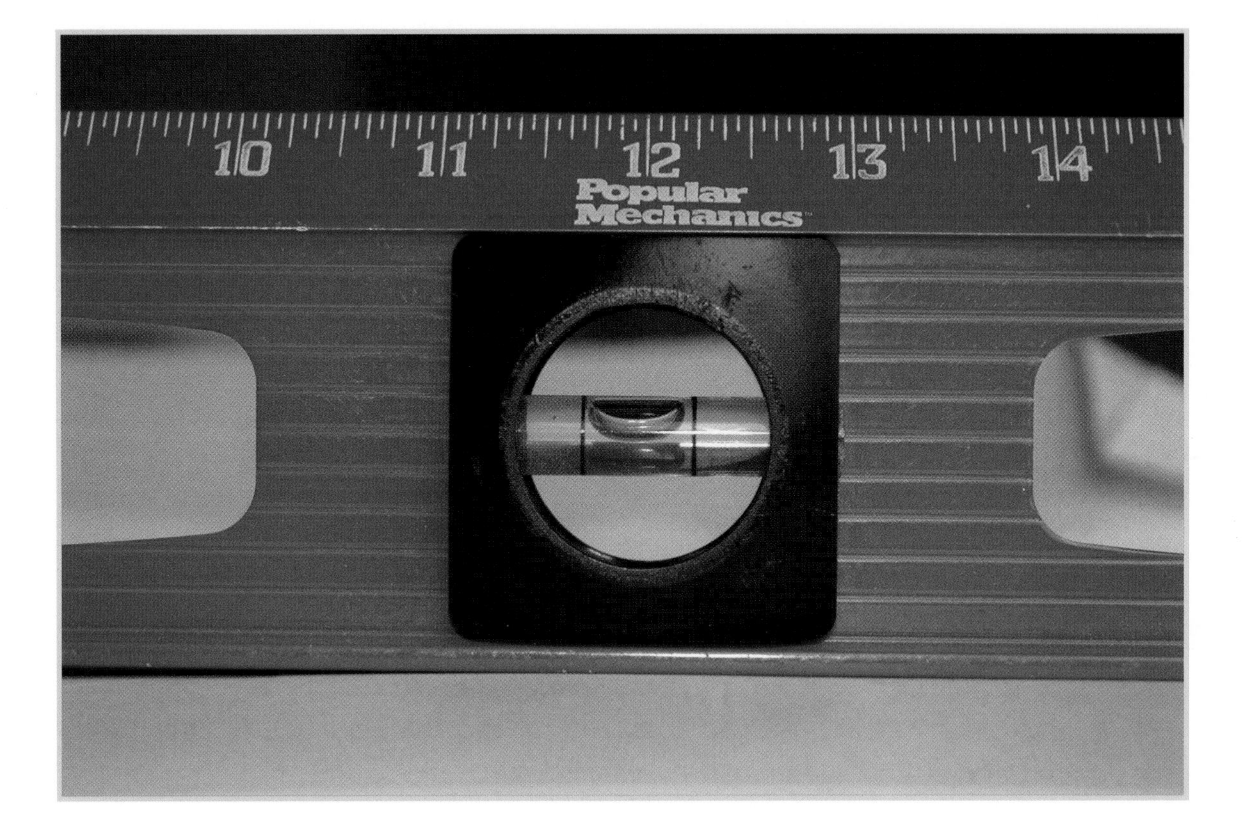

A carpenter puts a spirit level on a shelf. The bubble is in the middle of the tube. That means the shelf is level.

Is It Straight?
Chalk Lines and Plumb Bobs

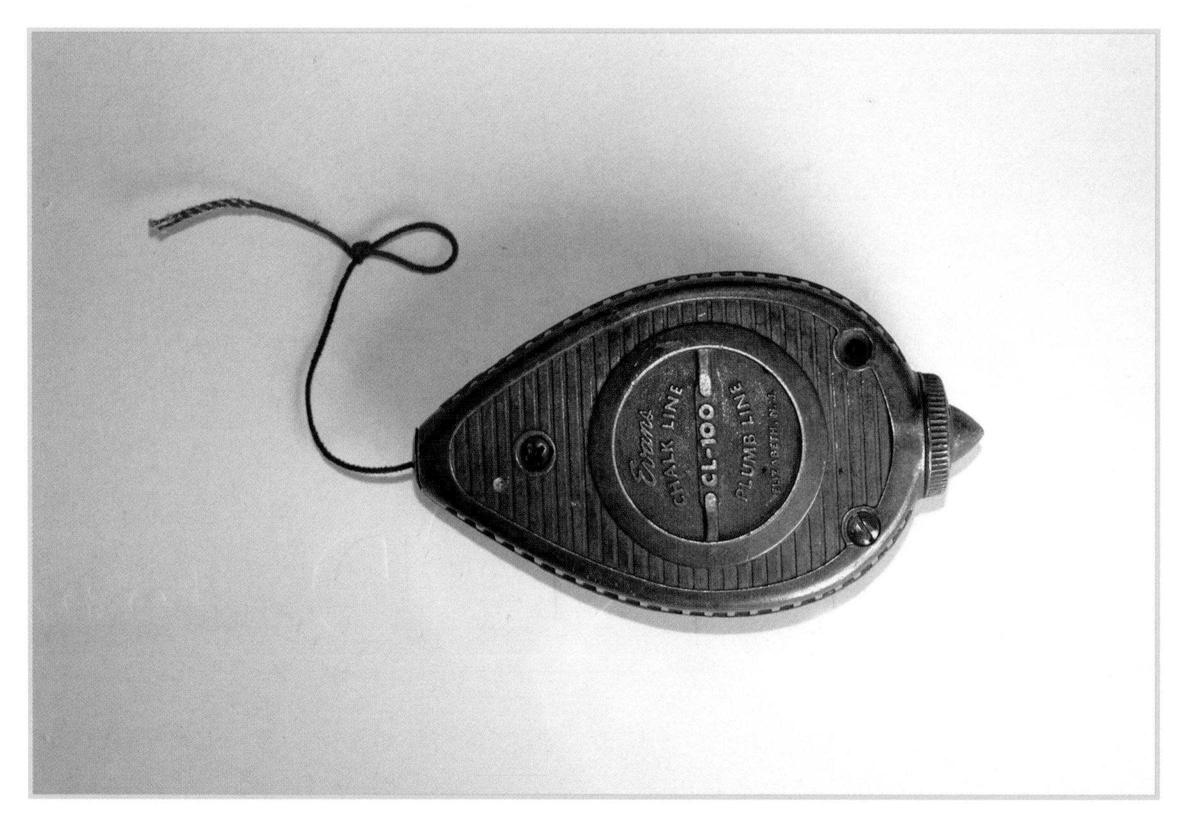

This string has chalk on it. A carpenter unwinds the string, then snaps it. The chalk makes a straight line on the wall.

This plumb bob is a string with a weight on it. It helps this worker make sure the door is straight up and down.

How Heavy? Balances and Scales

This is a balance. Put one thing in each bucket. It shows which thing is heavier.

A scale shows how much things **weigh**. This scale measures weight in pounds.

Measuring Distance

This is a laser level. The worker uses it to measure how far and how high something needs to be.

Another worker holds the grade pole. A grade pole is like a measuring stick. It works with the laser level.

Tool Safety

You should have an adult help you with any tools. You should wear gloves and goggles to protect yourself.

GLOSSARY

carpenter (KAHR puhn tur) — person who builds with wood

distance (DISS tuhnss) — how far

length (LENGKTH) — how long

level (LEV uhl) — a tool to help make things flat

tool (TOOL) — something that helps people do work

weigh (WAY) — how heavy

INDEX

FURTHER READING

Gresko, Marcia. *Measuring*. Gareth Stevens, 2004.
Schwartz, David M. *Millions to Measure.* HarperCollins: New York, 2003.

WEBSITES TO VISIT

www.thewoodcrafter.net/jr.html
www.enchantedlearning.com/dictionarysubjects/tools.shtml
www.bobthebuilder.com/usa/index.html

ABOUT THE AUTHOR

Patty Whitehouse has been a teacher for 17 years. She is currently a Lead Science teacher in Chicago, where she lives with her husband and two teenage children. She enjoys reading, gardening, and writing about science for children.